ROLE OF DIPEPTIDYL - PEPTIDASE IV (DPP-IV) IN TYPE II DIABETES

Allenki Venkatesham*

Dept of Pharmacology and Clinical Pharmacy,

SVS Group of Institutions, School of Pharmacy, Warangal – 506 009, AP, INDIA.

*Correspondence Address

Dr. Allenki Venkatesham
             M.Pharm, Ph.D

Professor, Dept of Pharmacology and Clinical Pharmacy,
SVS Group of Institutions, School of Pharmacy
Ramaram, Warangal
Office: 0870-6560834
Mobile: +91-9248022625
Fax: +91-870-2453900
E-mail: venkatkuc@gmail.com

TABLE OF CONTENTS

INTRODUCTION

Dipeptidyl-peptidase IV (DPP-IV) plays a great role in the scientific, pharmaceutical, and medical research. Dipeptidyl peptidase IV (CD26; E.C. 3.4.14.5) is plasma membrane glycoprotein exopeptidase that belongs to the prolyl oligopeptidase family (1,2). DP IV (EC 3.4.14.5, also DP 4, CD26, DPP IV, DPP-IV or DPP-4) was discovered in the 1960s as an amino peptidase (3) shown in figure1.

Figure 1 :Structure of DPP-IV

In the 1970s, the enzyme served initially as a model protein for the study of the catalytic mechanism of serine peptidases and for the investigation of the specifics of proline peptide bonds (4). In the 1980s, the potential of the enzyme to convert bioactive peptides *in vitro* was discovered, which intensified the search for its function *in vivo* (5). The early 1990s is characterized by numerous studies on the role of DP IV in immune responses, in particular T cell activation, signal transduction and T cell proliferation (6,7). DPP- IV was identified as characteristic antigen marker CD 26 of T cell activation and it was demonstrated that the protein is a component of the T cell receptor complex. Different binding partners were found within this context, for example adenosine deaminase (ADA) (8), the HIV mantle protein gp120 (9), tyrosine phosphatase CD 45 (10), fibronectin (11) and the renal sodium proton antiporter NHE3 (12).In the middle of the 1990s, the involvement of DPP IV in the metabolism and the regulation of the cytokines, chemokines and different peptide hormones triggered programs led to the development of DPPIV inhibitors (13). It was the discovery for the treatment of that most important of   metabolic diseases, type 2 diabetes, and led to the first patent application for the use of DPP- IV inhibition in the reduction of blood glucose (14).The DPP-IV three-dimensional structures help not only in understanding the unusual protease function of the enzyme, but also explain additional properties which differentiate the enzyme from the ''trypsin'' serine proteases and characterize it as an important physiological communication molecule.

STRUCTURE AND EXPRESSION

DPP IV is a serine-type peptidase and was first isolated from rat liver in 1966 (3).It is known as the lymphocyte cell surface protein CD26, widely expressed glycoprotein, that exhibits three principal biological activities in humans, it functions as an adenosine deaminase (ADA)-binding protein, it contributes to extracellular matrix binding and it exhibits post proline or alanine peptidase activity, thereby inactivating or in some cases generating biologically active peptides via cleavage at the N-terminal region after X-proline or X-alanine represented in table- 1 (15,16)

	Activity in serum (U/L)			Activity in citrated in plasma (U/L)		
	Mean	SD	Range	Mean	SD	Range
Women	27.5	6.1	17.0-50.8	23.2	4.4	14.0-39.7
Men	32.3	6.4	17.7-52.6	25.9	5.1	12.5-42.0

Table.1: Normal Human Dipeptidyl-Peptidase Activity

DP IV is a membrane-bound, homodimeric class II protein with a molecular weight of 110–150 kD per subunit. It is bound to the membrane by a transmembrane sequence of ca. 22 amino acids. The 6 cytosolic amino acids play no role in the binding functions. Different structural regions of the protein are assigned to the primary sequence shown in Figure 2.

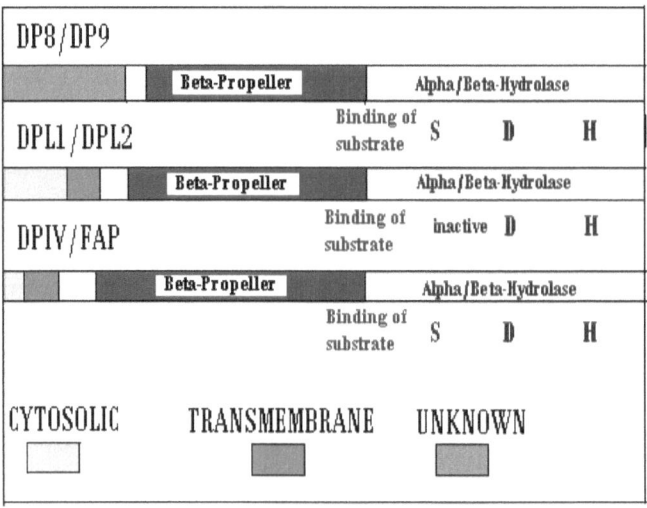

Figure 2. : Human DPP-IV gene family

Whereas the catalytically active residues of the enzyme are localized in the region of the α/β -hydrolase domain, the protein–protein interactions described occur mainly with region of the so-called β-propeller domain. The DPIV gene family has six members, including FAP, DP8, DP9 and the two nonenzymes DPL1 and DPL2. The human DPP-IV gene contains 26 exons, is localized to the long arm of chromosome 2 and intriguingly, is localized adjacent to the proglucagon gene which encodes GLP-1 and GLP-2, principal substrates for DPP-IV. The human DPP-IV cDNA encodes a predicted protein of 766 amino acids, with 6 amino acids in the cytoplasm, 22 residues spanning the plasma membrane and 738 amino acids comprising the extracellular domain. The cellular localization of DPP IV-like enzymes is schematically shown in figure 3.

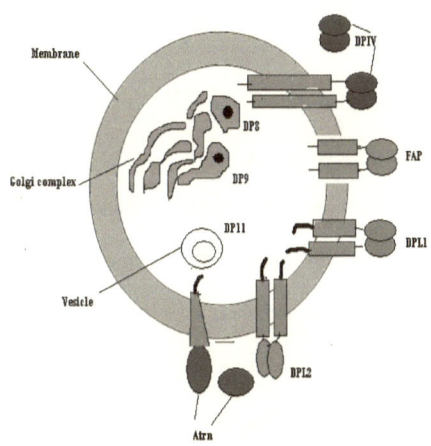

Figure 3: The cellular localization of DPP-IV like enzymes

DPP-IV is expressed in all organs, by capillary endothelial cells and activated lymphocytes and on apical surfaces of epithelial, including acinar, cells. In humans, DPP-IV is present in the gastrointestinal tract, biliary tract, exocrine pancreas, kidney, thymus, uterus, placenta, prostate, adrenal, parotid, sweat, salivary and mammary glands and endothelia of all organs examined, including liver, spleen, lungs and brain (17-22). Moderate amounts of sDPP IV are detected in serum/plasma (23-26). A high DPP IV activity is present in seminal fluid (27-29).

CLINICAL CHEMISTRY

a) Enzymatic assays

DPP IV activity in biological samples is determined by means of chromogenic or fluorogenic substrates. For example, the release of *para*-nitroaniline from Gly-Pro-*para*-nitroanilide is monitored spectrophotometrically as a function of time. This assay can easily be automated for use in microtiterplates or clinical analyzers. In our laboratory, 1 unit of activity is defined as the amount of enzyme that consumes 1 micromole of substrate per min at 37°C (assay conditions: 0.5 mM Gly-Pro-p-NA in 50 mM TRIS-HCl pH 7.6). Reference values were reported for serum and plasma of a relevant group of healthy adults (26). The activities in serum and citrated plasma were similar, considering the dilution caused by the addition of citrate. The DPP IV activity is lower in women than in men and tends to decrease with age represented in table 2. Apart from serum and plasma, DPP -IV activity has been measured in various biological fluids (urine, seminal plasma, amniotic fluid, synovial fluid) and cell suspensions. Soluble DPP IV/CD26 may originate from endothelial or epithelial cells and from circulating leukocytes. Serum DPP IV activity is invariably increased in patients with cirrhosis of the liver and disruptive hepatic disorders (30-32).In healthy individuals, the electrophoretic mobility of normal serum DPP IV coincides with that of lymphocyte DPP IV, but not with DPP IV purified from liver or kidney (33).Furthermore, plasma DPP IV in HIV-1-infected individuals is predominantly present in a sialylated state, corresponding to the form found on T lymphocytes (34). Immune reconstitution after antiretroviral therapy increases not only serum DPP-IV activity but also the number of CD26+CD4+ T cells (35-36). Low DPP-IV levels occur concurrently with an impaired immune status. However, serum DPP-IV activity does not always correlate with cytometric data and other immunological parameters. For example, in patients receiving immunosuppressant therapy after kidney transplantation, DPP-IV/CD26 expressed on the surface of PBL was transiently reduced during approximately 14 days after the operation, while serum activity was significantly reduced for more than 4 months (37). DPP IV levels have been reported as clinically relevant for various diseases that can be subdivided into four distinct categories: (1) solid tumors, (2) hematological malignancies, (3) auto immune and inflammatory diseases, and (4) infectious diseases such as AIDS and Hepatitis C.

b) DPP IV expression in cancer

DPP IV is expressed on many types of epithelial cells, it is often deregulated in human malignancies. The alteration of DPP IV membrane expression has been reported in several carcinoma (38). Loss of DPP IV expression is highly correlated with malignant transformation in melanocytes. There is an inverse relationship between metastatic capability of prostate cancer cells and DPP IV expressed on their surface (39-40).DPP-IV was proposed as a specific marker for differentiated thyroid carcinoma, whereas benign thyroid tissue is DPP IV-negative (41-42).Low-soluble DPPIV activity in serum is considered to have predictive value for colorectal cancer (43).

c) DPP IV expression in hematological malignancies

In healthy subjects, the high-density expression of DPP-IV is restricted to a particular set of memory T cells. Abnormal expression of DPP-IV/CD26 in hematological malignancies is related to the lineage of the transformed cells. DPP-IV/CD26 is considered a marker for certain aggressive T-cell lymphomas. (44). In mycosis fungoides, increase in CD4+ T cells concomitant with an elevation in the CD4/CD8 ratio and dim or absent surface DPP IV/CD26. This pattern distinguishes the disease from other disorders such as inflammatory dermatitis and atypical cutaneous lymphoid lesions (45-46).DPPIV/CD26 expression is also decreased in adult T-cell leukemia/lymphoma, T-cell acute lymphoblastic leukemia, and T-cell chronic lymphocytic leukemia (47). It is increased in anaplastic large cell lymphoma, precursor T-lymphoblastic lymphoma, hepatosplenic lymphoma, and B-cell chronic lymphocytic leukemia (48).

d) DPP IV expression in autoimmune diseases, inflammation and psychoneuro endocrine disorders

Several groups investigated DPP IV/CD26 as a putative marker for a specific T-cell response (T_H1 or T_H2) in (auto immune and inflammatory diseases). DPP IV/CD26 is over expressed on peripheral blood T cells in rheumatoid arthritis, while soluble DPP IV is significantly reduced, and this reduction is related to disease activity (49). Reduced serum activity was also observed in patients with inflammatory bowel disease (Crohns disease and ulcerative colitis). These patients have a higher number of CD26-positive cells coexpressing CD25 and a higher surface density of CD26. Elevated DPP IV activity in patients with anorexia nervosa and bulimia nervosa (50). DPP IV was increased in patients with atopic dermatitis (51). DPP IV levels were reduced in middle-aged and elderly obese patients with diabetes (52).

e) DPP IV expression during bacterial and viral Infections

During systemic bacterial infections and multi organ dysfunction, very high concentrations of the propeptide of calcitonin, procalcitonin (PCT) are found in the blood circulation without an increase of calcitonin levels. The penultimate proline of PCT (1-116) is well conserved, and only the truncated PCT(3-116) is found in the serum of patients with severe sepsis (53). A significant decrease in DPP IV activity was observed in patients with severe sepsis in relationship to the increase of procalcitonin.

II. FUNCTIONS

The roles of DPP IV in the immune system and tumor invasion appear to involve both enzymatic and nonenzymatic actions. Studies with DPP IV-negative animals and *in vivo* inhibition experiments contribute to the insight into the physiological role of this protein.

A. Role of DPP IV in Diabetes

Role of DPP IV in diabetes is based on the observation that DPP IV inactivates a number of peptides involved in glucose homeostasis as follows.

1. Glucagon like peptide-1

Glucagon like peptide -1(GLP-1) is a 30 amino acid peptide. Two forms of GLP-1 secreted by the L- cells of small intestine in response to meal ingestion, GLP-1(7-37) and GLP-1(7-36) amide differ by single amino acid. Both peptides are equipotent and exhibit identical plasma half lives and biological activities acting through same receptor (54 - 55), however majority (~80%) of circulating active GLP-1 appears to be GLP-1(7-36) amide (56). It amplifies meal-induced insulin release and synthesis in a glucose-dependent manner. GLP-1 directly stimulates glucose dependent insulin secretion via an increase in beta cell cAMP (57), through both protein kinase A-dependent and independent mechanisms, with activation of signaling through G-proteins contributing to control of insulin exocytosis (58). GLP-1 receptor (GLP-1R) activation also promotes calcium mobilization (59) and closure of the ATP sensitive K$_{ATP}$ channel (60). It inhibits glucagon secretion. (61), stimulates insulin biosynthesis as well as insulin gene transcription (62), slows gastric emptying (63-64), reduced food intake (65) and promotes satiety in humans (66).

2. Gastric inhibitory peptide.

GIP (42 amino acids) is secreted by the endocrine K cells of the proximal intestine in response to nutrients, especially fats (67). Originally called gastric inhibitory peptide, it was renamed glucose-dependent insulinotropic polypeptide to reflect its most important function as a gut hormone. GIP acts through a G protein-coupled receptor in a large array of tissues (68). GIP stimulates insulin release from pancreatic β □cells in the presence of elevated glucose levels. As opposed to GLP-1, GIP stimulates the release of glucagon. The truncation of GIP by DPP IV, resulting in GIP (3-42), has been shown *in vitro* and *in vivo* (69). The possible influence of DPP IV-mediated cleavage of these peptides on the stimulation of insulin secretion in cells is schematically shown in figure.4.

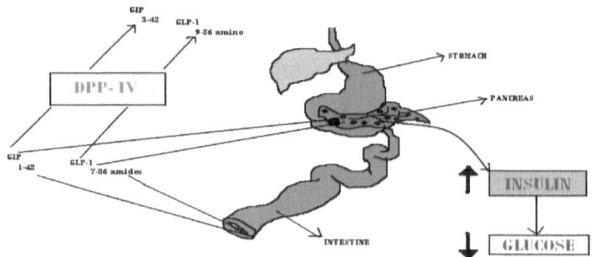

Figure.4 : Role of DPP-IV, GLP-1, GIP in glucose homeostasis.

3. Gastrin-releasing peptide

Gastrin-releasing peptide (GRP) is a member of the bombesin family of peptide hormones. GRP is produced in the brain, the intrinsic neurons of the gut, and the parasympathetic neurons of the pancreas (70).The GRP receptor belongs to the G protein-coupled receptor family and is present on the epithelial cells lining the gastric antrum and in the pancreas. The insulinotropic action of GRP is due to direct stimulation of the □cells, to activation of postganglionic parasympathetic nerves at the ganglionic level, and to stimulation of GLP-1 release (71). Human GRP is an excellent substrate for DPP IV, which sequentially removes two dipeptides with almost equal efficiency. The selectivity constant for GRP is significantly higher than for GLP-1 and GIP, whose active intact peptide levels are regulated by DPP IV.

4. Several (Neuro) peptides interact with insulin-releasing cells

Several peptides that are involved in the regulation of endocrine pancreatic secretion originate from the gut. They form the enter insular axis, the signaling pathways between gut and pancreatic islets that enhance the insulin response to absorbed nutrients. GLP-1 and GIP are the most important insulin releasing hormones (incretins) in this axis.

4. Prolongation of incretin action improves glucose tolerance in type 2 diabetes

The *in vivo* removal of N-terminal dipeptides from incretins and other regulatory peptides such as glucagon has been demonstrated with N-terminus specific antibodies and by liquid chromatography-mass spectrometry (72). The plasma levels of GLP-1 transiently rise within 5 to 15 min after a meal. The *in vivo* half-life of intact GLP-1 in humans is 1 to 2 min (73). GLP-1 administration to type 2 diabetes patients resulted in an improved glucose tolerance and decreased food intake (74).

5. In vivo effect of DPP IV inhibition on glucose tolerance

In earlier studies it was reported that that improvement of glucose tolerance by *in vivo* inhibition of DPP IV. This effect holds true in several species (mice, rats, dogs, pigs). After glucose challenge, the lack of DPP IV expression leads to significantly faster blood glucose clearance, accompanied with increased insulin levels. No significant differences are detected in the fasting levels of glucose, insulin, or GLP-1 in these mice compared with DPP IV-positive animals.

B. Processing of Bioactive Peptides

1. Human Bioactive Peptides as DPP IV Substrates A number of bioactive peptides have the N-terminal X-Pro or X-Ala required for cleavage by DPP IV. The presence of a proline near the amino terminus serves as a structural protection against nonspecific proteolytic degradation.

2. Substance P

Substance P is a widespread neuropeptide that originates from sensory neurons and signals by activating heptahelical G protein-coupled receptors. Together with other tachykinins, the peptide is responsible for nociceptive transmission from the peripheral to the central nervous system (75).It also has a well-established role in immunity. Substance P induces the release of inflammatory mediators from mast cells and causes an increase in vascular permeability. Substance P is inactivated in the blood circulation by DPP IV and angiotensin converting enzyme (76).

3. Casomorphin

The heptapeptide casomorphin is formed in the intestine when the milk protein casein is not properly digested. Casomorphin has potent opioid activity and therefore is referred to as an exorphin. DPP IV, present on the brush-border membrane of the intestine, first converts the peptide into the active casomorphin (3-7), which is quickly processed further by DPP IV into inactive fragments (77)

4. The analgesic brain peptides

A number of analgesic tetra peptides that are found in the brain contain a penultimate proline, and central inhibition of DPP IV-activity induces analgesia (78).Endomorphin-2 is abundantly present in the cortex of the human brain and has a high affinity for the μ-opioid receptors that produce analgesia. The *in vivo* inactivation of endomorphin-2 by DPP IV was demonstrated in rats. Central inhibition of DPP IV enforced the analgesic actions of exogenous endorphin 2 (79).

5. The PACAP/glucagon family of peptides

GLP-2, secretin, vasoactive intestinal peptide (VIP), pituitary adenylate cyclase-activating peptide (PACAP), glucose-dependent insulinotropic peptide (GIP, also referred to as gastric inhibitory peptide), growth hormone-releasing factor (GRF), and peptide histidine methionine (PHM). These peptides are related in the structure of their N-terminal region, distribution, function, and receptors (80). In this family, the intact N-terminus (Tyr-Ala, His-Ala, or His-Ser) is necessary for biological activity, and truncation by DPP IV causes inactivation. Since the PACAP/glucagon family members with a penultimate serine were considered to be 'DPP IV-resistant', the truncation of glucagon by DPP IV was unexpected (81).

CONCLUSION

The discovery of the regulatory function of the enzyme DP IV in glucose homeostasis in the middle of the 1990s has allowed the development of novel oral antidiabetics. In contrast to former diabetic pharmaceuticals, this therapy is causal and pleiotropic since it utilizes multiple functions, notably the small intestine hormones GLP-1 and GIP. Many bioactive peptides qualify to be DPP IV substrates. An improved insight into the structural and enzymological aspects of DPP IV is also a prerequisite for the development of more selective inhibitors. Experiments with animals and the first clinical studies in men have firmly established DPP IV as a drug target for the treatment of type 2 diabetes. Following advancements in functional genomics and proteomics, scientists realized that proteins do not function alone but are interconnected in networks. Considering the number of protein-binding sites on DPP IV and its association with regulated processes such as T cell activation, cell migration, and invasiveness, we can only conclude that DPP IV is tied in with several of these networks.

REFERENCES

1. Lambeir A-M, Durinx C, Scharpe´ S, De Meester I. Dipeptidyl- peptidase IV from bench to bedside: an update on structural properties,functions and clinical aspects of the enzyme DPP IV. Crit Rev ClinLab Sci 2003;40: 209–94.

2. De Meester I, Lambeir A-M, Proost P, Scharpe´ S. Dipeptidyl peptidase IV substrates Adv Exp Med Biol 2003;524:3-13.

3. V.K. Hopsu-Havu, G.G. Glenner, A new dipeptide naphthylamidase hydrolyzing glycyl-prolyl-beta-naphthylamide, Histochemie, 1966;   7:197-201.

4. H.-U. Demuth, J. Heins, On the catalytic mechanism of dipeptidyl peptidase IV, in: B. Fleischer (Ed.), Dipeptidyl Peptidase IV (CD26) in Metabolism and the Immune Response, R.G. Landes Company, Austin, 1995, pp. 1-35.

5. A. Yaron, The role of proline in the proteolytic regulation biologically active peptides, Biopolymers 1987;26: S215-S222 (Suppl.).

6. B. Fleischer, CD26: a surface protease involved in T-cell activation,   Immunol. Today 1994; 15: 180-184.

7. A.J. Ulmer, T. Mattern, H.D. Flad, Expression of CD26 (dipeptidyl   peptidase IV) on memory and naive T lymphocytes, Scand. J. Immunol. 1992 ; 35: 551– 559.

8. J. Kameoka, T. Tanaka, Y. Nojima, S.F. Schlossman, C. Morimoto,Direct association of adenosine deaminase with a T cell activation antigen, CD26, Science 1993; 261: 466–469

9. C. Callebaut, B. Krust, E. Jacotot, A.G. Hovanessian, T cell activation antigen, CD26, as a cofactor for entry of HIV in CD4+ cells, Science 1993; 262 : 2045– 2050.

10. Y. Torimoto, N.H. Dang, E. Vivier, T. Tanaka, S.F. Schlossman, C.Morimoto, Coassociation of CD26 (dipeptidyl peptidase IV) with CD45 on the surface of human T lymphocytes, J. Immunol. 1991; 147: 2514– 2517.

11. C. Hanski, T. Huhle, W. Reutter, Involvement of plasma membrane dipeptidyl peptidase IV in fibronectin-mediated adhesion of cells on collagen, Biol.   Chem. Hoppe-Seyler 1985; 366: 1169– 1176.

12. A.C.C. Girardi, B.C. Degray, T. Nagy, D. Biemesderfer, P.S. Aronson, Association of Na+–H+ exchanger isoform NHE3 and dipeptidyl peptidase TV in the renal proximal tubule, J. Biol. Chem. 2001; 276: 46671– 46677.

13. R. Mentlein, B. Gallwitz, W.E. Schmidt, Dipeptidyl-peptidase IV hydrolyses gastric inhibitory polypeptide, glucagon-like peptide-1(7-36)amide, peptide histidine methionine and is responsible for their degradation in human serum, Eur. J. Biochem.1993;214: 829–835.

14. R.P. Pauly, H.-U. Demuth, F. Rosche, J. Schmidt, H.A. White, C.H.S. McIntosh, R.A. Pederson, Inhibition of dipeptidyl peptidase IV (DPPIV) in rat results in improved glucose tolerance, Regul. Pept.1996; 64: 148.

15. Mentlein R, Dipeptidyl peptidase IV (CD26) role in the activation of regulatory peptides. Regul.Pept.1999; 85: 9-24.

16. Gorreel MD, Gysbers V, Mccaughan GW, CD26: a multifunctional integral membrane and secreted protein of activated lymphocytes. Scand J. Immunol.2001; 54:249-264.

17. McCaughan, G.W., Gorrell, M. D., Bishop, G. A. et al. Molecular pathogenesis of liver disease:an approach to hepatic inflammation, cirrhosis and liver transplant tolerance. Immunol. Rev.2000; 174, 172–191

18. Durinx, C., Lambeir, A. M., and Bosmans, E. et al..Molecular characterization of dipeptidyl peptidase activity in serum: soluble CD26/dipeptidyl peptidase IV is responsible for the release of X-Pro dipeptides. Eur. J. Biochem. 2000; 267, 5608-5613

19. Rasmussen, H. B., Branner, S., Wiberg, F. C. and Wagtmann, N. Crystal structure of human DPP-IV/CD26 in complex with a substrate analogue. Nat. Struct. Biol.2003; 10, 19–25

20. Abbott, C. A., McCaughan, G. W. and Gorrell, M. D. Two highly conserved glutamic acid residues in the predicted β propeller domain of dipeptidyl peptidase IVare required for its enzyme activity. FEBS Lett.1999; 458, 278–284

21. Ajami, K., Abbott, C. A., Obradovic, M. et al. (2003) Structural requirements for catalysis, expression and dimerisation in the CD26/DPIV gene family. Biochemistry 2003; 42, 694–701

22. Gorrell, M. D., Gysbers, V. and McCaughan, G.W. CD26: a multifunctional integral membrane and secreted protein of activated lymphocytes. Scand. J. Immunol. 2001; 54, 249–264

23. Iwaki-Egawa S, Watanabe Y, Kikuya Y, et al. Dipeptidyl peptidase IV from human serum: purification, characterization, and N-terminal amino acid sequence. J Biochem (Tokyo) 1998; 124: 428–433.

24. Hino M, Nagatsu T, Kakumu S, *et al.* Glycylprolyl beta-naphthyl amidase activity in human serum. *Clin Chim Acta* 1975;62:5–11.

25. Hino M, Fuyamada H, Hayakawa T, *et al.* X-Prolyl dipeptidyl- aminopeptidase activity, with X-proline *p*-nitroanilides as substrates, in normal and pathological human sera. *Clin Chem* 1976; 22: 1256–1261.

26. Durinx C, Neels H, Van der Auwera JC, *et al.* Reference values for plasma dipeptidylpeptidase IV activity and their association with other laboratory parameters. *Clin Chem Lab Med* 2001; 39: 155–159.

27. Gossrau R. Peptidases II. Localization of dipeptidylpeptidase IV (DPP IV). Histochemical and biochemical study. *Histochemistry* 1979; 60: 231–48.

28. Vanhoof G, De Meester I, van Sande M, *et al.* Distribution of proline-specific aminopeptidases in human tissues and body fluids. *Eur J Clin Chem Clin Biochem* 1992; 30: 333–338.

29. Wilson MJ, Ruhland AR, Pryor JL, *et al.* Prostate specific origin of dipeptidyl Peptidase IV (CD-26) in human seminal plasma. *J Urol* 1998; 160: 1905–1909

30. Perner F, Gyuris T, Rakoczy G, *et al.* Dipeptidyl peptidase activity of CD26 in serum and urine as a marker of cholestasis: experimental and clinical evidence. *J Lab Clin Med* 1999;134: 56–67.

31. Senten K, Van der Veken P, Bal G, *et al.* Development of potent and selective dipeptidyl peptidase II inhibitors. *Bioorg Med Chem Lett* 2002; 12: 2825.

32. Lakatos PL, Firneisz G, Rakoczy G, *et al.* Elevated serum dipeptidyl peptidase IV (CD26, EC 3.4.14.5) activity in patients with primary biliary cirrhosis. *J Hepatol* . 1999; 30: 740.

33. Kasahara Y, Fujii N, Mizukoshi M, *et al.* Multiple forms of glycylprolyl dipeptidylaminopeptidase in serum and tissues. *Jpn J Clin Chem* 1983; 12: 89–93.

34. Smith RE, Talhouk JW, Brown EE, *et al.* The significance of hypersialylation of dipeptidyl peptidase IV (CD26) in the inhibition of its activity by Tat and other cationic peptides. CD26: a subverted adhesion molecule for HIV peptide binding. *AIDS Res Hum Retroviruses* 1998; 14: 851–868.

35. Zanussi S, Simonelli C, Bortolin MT, *et al.* Immunological changes in peripheral blood and in lymphoid tissue after treatment of HIV-infected subjects with highly active anti-retroviral therapy (HAART) or HAART + IL-2. *Clin Exp Immunol.*1999; 116: 486–492.

36. Keane NM, Price P, Lee S, *et al.* An evaluation of serum soluble CD30 levels and serum CD26 (DPP IV) enzyme activity as markers of type 2 and type 1 cytokines in HIV patients receiving highly active antiretroviral therapy. *Clin Exp Immunol* 2001; 126: 111–116.

37. Korom S, De Meester I, Stein A, *et al.* Das T-Zell-Antigen CD26/DPP IV als Marker der Immunomodulation in humanen Empfängern allogener Nierentransplantate. *Transplantationsmedizin* 2000; 2000: 72.

38. Sakamoto J, Watanabe T, Teramukai S, *et al.* Distribution of adenosine deaminase binding protein in normal and malignant tissues of the gastrointestinal tract studied by monoclonal antibodies. *J Surg Oncol* 1993; 52: 124–134.

39. Aratake Y, Kotani T, Tamura K, *et al.* Dipeptidyl aminopeptidase IV staining of cytologic preparations to distinguish benign from malignant thyroid diseases. *Am J Clin Pathol* 1991; 96: 306–310.

40. Tanaka T, Umeki K, Yamamoto I, *et al.* CD26 (dipeptidyl peptidase IV/DPP IV) as a novel molecular marker for differentiated thyroid carcinoma. *Int J Cancer* 1995 ; 64: 326–331.

41. Umeki K, Tanaka T, Yamamoto I, *et al.* Differential expression of dipeptidyl peptidase IV (CD26) and thyroid peroxidase in neoplastic thyroid tissues. *Endocr J* 1996; 43: 53–60.

42. Aratake Y, Umeki K, Kiyoyama K, *et al.* Diagnostic utility of galectin-and CD26/DPP IV as preoperative diagnostic markers for thyroid nodules. *Diagn Cytopathol* 2002; 26: 366–372.

43. Cordero OJ, Ayude D, Nogueira M, *et al.* Preoperative serum CD26 levels: diagnostic efficiency and predictive value for colorectal cancer. *Br J Cancer* 2000; 83: 1139–1146.

44. Verstovsek S, Cabanillas F, Dang NH. CD26 in T-cell lymphomas: a potential clinical role? *Oncology (Huntingt)* 2000; 14: 17–23.

45. Jones D, Dang NH, Duvic M, *et al.* Absence of CD26 expression is a useful marker for diagnosis of T-cell lymphoma in peripheral blood. *Am J Clin Pathol* 2001; 115: 885–892.

46. Bernengo MG, Novelli M, Quaglino P, *et al.* The relevance of the CD4+ CD26-subset in the identification of circulating Sézary cells. *Br J Dermatol* 2001; 144: 125–135.

47. Kondo S, Kotani T, Tamura K, *et al.* Expression of CD26/dipeptidyl peptidase IV in adult T cell leukemia/lymphoma (ATLL). *Leuk Res* 1996; 20: 357–363.

48. Bauvois B, De Meester I, Dumont J, *et al.* Constitutive expression of CD26/ dipeptidylpeptidase IV on peripheral blood B lymphocytes of patients with B chronic lymphocytic leukaemia. *Br J Cancer* 1999; 79:1042–1048.

49. Cordero OJ, Salgado FJ, Mera-Varela A, *et al.* Serum interleukin-12, interleukin-15, soluble CD26, and adenosine deaminase in patients with rheumatoid arthritis. *Rheumatol Int* 2001; 21: 69–74.

50. Hildebrandt M, Rose M, Monnikes H, *et al.* Eating disorders: a role for dipeptidyl peptidase IV in nutritional control. *Nutrition* 2001; 17:451- 454.

51. Katoh N, Hirano S, Suehiro M, *et al.* Soluble CD30 is more relevant to disease activity of atopic dermatitis than soluble CD26. *Clin Exp Immunol* 2000; 121: 187-192.

52. Meneilly GS, Demuth HU, McIntosh CH, *et al.* Effect of ageing and diabetes on glucosedependent insulinotropic polypeptide and dipeptidyl peptidase IV responses to oral glucose. *Diabet Med* 2000; 17: 346–350.

53 . Bergmann A, Bohuon C. Decrease of serum dipeptidyl peptidase activity in severe sepsis patients: relationship to procalcitonin. *Clin Chim Acta* 2002; 3211: 123–126.

54 . Orskov C,Wettergren A,Holst JJ. Biological effects and metabolic rates of glucose like peptide -1(7-36) amide and glucagon like peptide 1(9-37) in healthy subjects are indistinguishable. *Diabetes*1993; 42: 658-661.

55 . Nauck MA , Kleine N, Orskov C, Holst JJ,Willms B, Creutzfeldt W. Normalisation of fasting hyperglycemia by exogenous glucagon like peptide 1(7-36 amide) in type 2 diabetic patients. Diabetolgia 1993; 36: 741-744.

56.   Orskov C,Rabenhoj L,Wettergren A, Kofold H,Holst JJ: Tissue and plasma concentrations of amidated and glycine extended glucagon like peptide -1 in humans. 1994; 43: 535-539.

57.   Drucker dj, philippe j, mojsov s, Chick wl, habener jf: Glucagonlike peptide I stimulates insulin gene expression and increases cyclic AMP levels in a rat islet cell line. *Proc. Natl. Acad. Sci. USA* (1987) 84:3434-3438.

58.   Kashima Y, Miki T, Shibasaki T*et al.*: Critical role of cAMP-GEFII/Rim2 complex in incretin-potentiated insulinsecretion. *J. Biol. Chem.* 2001; 276: 46046-46053.

59. Holz Gg, Leech Ca, Habener JF:Activation of a cAMP-regulated Ca2+- signaling pathway in pancreatic b-cells bythe insulinotropic hormone glucagon-like peptide-1. *J. Biol. Chem.* 1995; 270:17749-17757.

60.   Light pe, Manning fox JE, Riedel MJ, and Wheeler MB: Glucagonlike peptide-1 inhibits pancreatic ATPsensitive potassium channels via a protein kinase A- and ADP-dependent mechanism. *Mol. Endocrinol.* 2002; 16:2135-2144.

61.   Toft – Nielsen MB, Damholt MB,Madasbad S,Hilsted LM, Hughes TE , Michelsen BK, Holst JJ: Determinants of the impaired secretion of glucagon like peptide -1 in type-2 diabetes patients. *J Clin Endocrinol Metab* 2001, 86: 3717-3723.

62. Fehmann HC, Hbener JF : Insulinotropic hormone glucagon like peptide-1 (7-37) stimulation of proinsulin gene expression and proinsulin biosynthesis in insulinoma beta TC-1 cells. *Endocrinology* 1992, 130: 159-166.

63. Wettergren A, Schjoldager b , Mortensen PE, Mythre J, Christiansen J, Holst JJ: Truncated GLP-1 (Proglucagon 78-107 - amide) inhibits gastric and pancreatic functions in man. *Dig Dis Sci* 1993 ; 38: 665-673.

64. Nauck MA, Niedereichholz U, Etter R, Holst JJ, Orskov C, Ritzel R ,Schimiegel WH Glucagon like peptide -1 inhibition of gastric emptying outweighs its insulintropic effects in healthy humans. *Am J Physiol* 1997; 273: E981-988.

65. Verdich C, Flint A,Gutzwiller JP, Naslund E , Begliner C, Hellstrom PM, Long SJ, Astrup A: A meta analysis of the effect of glucagon like peptide -1(7-36) amide in humans. *J Clin Endocrinol Metab* 2001; 86: 4382-4389.

66. Flint A, Raben A, Holst JJ: Glucagon like peptide -1(7-36) promotes satiety and suppression energy intake in humans. J Clin Invest. 1998; 101: 515-520.

67. Krarup T, Holst JJ, Larsen KL. Responses and molecular heterogeneity of IR-GIP after intraduodenal glucose and fat. *Am J Physiol* 1985; 249: E195–200.

68. Usdin TB, Mezey E, Button DC, *et al.* Gastric inhibitory polypeptide receptor, a member of the secretin-vasoactive intestinal peptide receptor family, is widely distributed in peripheral organs and the brain. *Endocrinology* 1993; 133:2861-2870

69. Jörnvall H, Carlquist M, Kwauk S, *et al.* Amino acid sequence and heterogeneity of gastric inhibitory polypeptide (GIP). *FEBS Lett* 1981; 123: 205–210.

70. Moghimzadeh E, Ekman R, Hakanson R, *et al.* Neuronal gastrin-releasing peptide in the mammalian gut and pancreas. *Neuroscience* 1983; 10: 553–563.

71. Karlsson S, Sundler F, Ahrén B. Insulin secretion by gastrin-releasing peptide in mice: ganglionic versus direct islet effect. *Am J Physiol* 1998; 274: E124–129.

72. Deacon CF, Wamberg S, Bie P, *et al.* Preservation of active incretin hormones by inhibition of dipeptidyl peptidase IV suppresses meal-induced incretin secretion in dogs. *J Endocrinol* 2002; 172: 355–362.

73. Deacon CF, Nauck MA, Toft-Nielsen M, *et al.* Both subcutaneously and intravenously administered glucagon-like peptide I are rapidly degraded from the NH2–terminus in type II diabetic patients and in healthy subjects. *Diabetes* 1995; 44: 1126–1131.

74. Toft-Nielsen MB, Madsbad S, Holst, JJ. Continuous subcutaneous infusion of glucagonlike peptide 1 lowers plasma glucose and reduces appetite in type 2 diabetic patients. *Diabetes Care* 1999;22:1137–1143.

75. Zubrzycka M, Janecka A. Substance P: transmitter of nociception (Minireview). *Endocr Regul* 2000; 34: 195–201.

76. Palmieri FE, Ward PE. Mesentery vascular metabolism of substance P. *Biochim Biophys Acta* 1983; 755: 522–525.

77. Hartrodt B, Neubert K, Fischer G, *et al.* Degradation of beta- casomorphin-5 by prolinespecific- endopeptidase (PSE) and post-proline-cleaving-enzyme (PPCE). Comparative studies of the beta-casomorphin-5 cleavage by dipeptidyl-peptidase IV. *Pharmazie* 1982; 37:72–83.

78. Ronai AZ, Timar J, Mako E, *et al.* Diprotin A, an inhibitor of dipeptidyl amino peptidase IV(EC 3.4.14.5) produces naloxone-reversible analgesia in rats. *Life Sci* 1999; 64: 145–152.

79. Shane R, Wilk S, Bodnar RJ. Modulation of endomorphin-2–induced analgesia by dipeptidyl peptidase IV. *Brain Res* 1999; 815: 278–286.

80. Sherwood NM, Krueckl SL, McRory JE. The origin and function of the pituitary adenylate cyclase-activating polypeptide (PACAP)/glucagon superfamily. *Endocr Rev* 2000; 21: 619–670.

81. Hinke SA, Pospisilik JA, Demuth HU, *et al.* Dipeptidyl peptidase IV (DPIV/CD26) degradation of glucagon. Characterization of glucagon degradation products and DPIVresistant analogs. *J Biol Chem* 2000; 275: 3827–3834.

www.ingramcontent.com/pod-product-compliance
Lightning Source LLC
Chambersburg PA
CBHW021858170526
45157CB00006B/2504